INTRODUCTION

Of the millions of stars in the sky, about 2,000 are visible to the naked eye on any given night. This number is far less in cities where streetlights tend to blot out duller stars. This guide highlights the brightest stars and constellations that are visible from cities in the Southern Hemisphere in summer and winter.

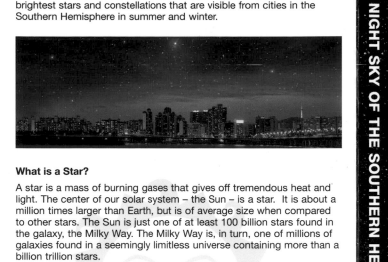

What is a Star?

A star is a mass of burning gases that gives off tremendous heat and light. The center of our solar system – the Sun – is a star. It is about a million times larger than Earth, but is of average size when compared to other stars. The Sun is just one of at least 100 billion stars found in the galaxy, the Milky Way. The Milky Way is, in turn, one of millions of galaxies found in a seemingly limitless universe containing more than a billion billion stars.

The stars in the night sky are simply the suns of different solar systems and galaxies. How bright they appear is affected by their size, how hot they burn and their distance from Earth.

Our sun, which is by far the brightest star in the day or night sky, is an average-sized star located about eight light minutes from Earth. By comparison, Rigel, a star in the constellation Orion, is estimated to be 74–75 times larger in diameter and 50,000 times more powerful than the Sun, but is merely a spark in the night sky since it is about 860 light-years from Earth. Light travels 186,000 miles (300,000 km) a second, 11.2 million miles (18 million km) a minute, 672 million miles (1,080 million km) an hour, 16 billion miles (26 billion km) a day and about 6 trillion miles (9.7 trillion km) a year.

This guide is intended to serve as a star "finder" and should supplement a field guide to astronomy that can provide detailed information about stars, constellations and celestial events. The star charts highlight the most prominent constellations and brightest stars visible from cities throughout the Southern Hemisphere.

Waterford Press produces reference guides that introduce novices to nature, science, survival and outdoor recreation. Product information is featured on the website: www.waterfordpress.com

Text and illustrations © 2001, 2021 by Waterford Press Inc. All rights reserved. Cover images © iStock Photo. To order, call 800-434-2555. For permissions, or to share comments, e-mail editor@waterfordpress.com. For information on custom-published products, call 800-434-2555 or e-mail info@waterfordpress.com.

978-1-62005-497-0 $7.95 U.S.

Made in the USA

The Night Sky of the Southern Hemisphere

Star Charts Glow in the Dark

A Folding Glow-in-the-Dark Pocket Guide to Prominent Stars & Constellations South of the Equator

---

CONSTELLATIONS

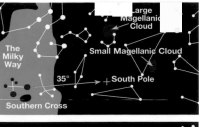

Prominent features of the southern night sky include the Milky Way, The Southern Cross and the Large and Small Magellanic Clouds.

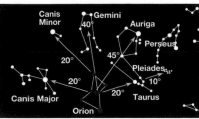

Orion is a prominent summer constellation in the southern hemisphere that can be used as a guide to other stars and constellations.

Andromeda, The Chained Princess
Her mother, Cassiopeia the Queen, was so boastful of Andromeda's beauty that the Sea God Neptune had her chained to a cliff and sacrificed to a sea monster.

Aquila, The Eagle
Aquila is marked by the bright star, Altair.

Aries, The Ram
The Golden Fleece of Aries was sought by Jason and his Argonauts.

Aquarius, The Water Bearer
A collection of dim stars found beneath Pegasus. A meteor shower occurs in its vicinity each May.

Argo Navis, The Ship Argo
The Argonauts were a group of Greek heroes (including Jason, Castor and Pollux) that sailed in search of the Golden Fleece. Their Ship Argo is composed of four constellations including Puppis (the stern), Carina (the keel), Pyxis (the compass) and Vela (the sails). The star Canopus in Carina is the second brightest star in the southern night sky.

Auriga, The Charioteer
A line from the star in the middle of Orion's belt between his shoulder stars for 45° leads to Capella, one of the brightest stars in the night sky.

Bootes, The Herdsman
Kite-shaped constellation is marked by the bright star Arcturus.

Canis Major, The Big Dog
Following the line of Orion's belt for 20° to the east leads to Sirius, the brightest star in the night sky. Also called the Dog Star, it marks Orion's big hunting dog.

Canis Minor, The Little Dog
Following the line of Orion's shoulder stars for 20° to the east leads to Procyon, the brightest star in his smaller dog.

Capricornus, The Sea Goat
The result of a botched transformation from a goat to a fish, the star pattern suggests bikini underpants to some.

Centaurus, The Centaur
Contains two bright stars which form "pointers" to the Southern Cross. The star Rigel Kentaurus is the closest star to our solar system, a mere 4.3 light-years away. Centaurus also contains one of the brightest clusters of stars in the sky – Omega Centauri.

Cetus, The Sea Monster
Located in the southern sky in summer, Cetus was a whale-like sea monster sent by Neptune to dispatch Andromeda.

Columba, The Dove
Commemorates the dove Noah sent from his ark to find dry land.

Corona Borealis, The Northern Crown
Small arc of bright stars is located near Bootes.

Corvus, The Crow
Box-shaped constellation found below Virgo.

Crux, The Southern Cross
Composed of five bright stars, Crux is near a dark area in the Milky Way devoid of stars called the "Coal Sack". The approximate location of the South Pole can be found by extending a line from the base of the cross for about 35° (about four times the length of the long axis of the cross from the base downwards).

Cygnus, The Swan
The bright star Deneb is at the "tail" of this constellation, which features an asterism known as the Northern Cross.

Delphinus, The Porpoise
Kite-like compact group of stars.

Dorado, The Fish Dolphin
Linear constellation contains the Large Magellanic Cloud.

Eridanus, The River
Long, winding chain of dim stars beginning near Orion.

Gemini, The Twins
Following a line from Orion's stars Rigel through Betelgeuse for 40° leads to the bright stars Castor and Pollux in Gemini.

Grus, The Crane
Symbol for the star-watcher in ancient Egypt.

Hercules, The Strongman
A line drawn from the star Vega to Arcturus passes through the box-like waist of Hercules.

Hydra, The Water Serpent
The largest constellation is composed of a chain of faint stars and a box-like head.

Hydrus, The Water Snake
Found between the Large and Small Magellanic Clouds.

Leo, The Lion
Leo is marked by the bright star Regulus.

Large and Small Magellanic Clouds
These "fuzzy" areas in the night sky are the two galaxies nearest our own (both are over 160,000 light-years away).

Lepus, The Hare
One of Orion's favorite game animals is found at his feet.

Lyra, The Lyre
Lyra is marked by the bright star, Vega.

Mensa, The Table
Found near the Large Magellanic Cloud.

The Milky Way
The path of our galaxy is visible in the night sky as a cloudy band of light. The Milky Way is much brighter in the southern hemisphere than in the north since the section of the galaxy visible from the south contains more stars.

Octans, The Octant
Constellation in the vicinity of the South Celestial Pole commemorates an important navigational tool of the 1700s.

Ophiuchus, The Serpent Bearer
The ship's doctor for the Argonauts of Greek legend was so skilled he could raise men from the dead.

Orion, The Hunter
Orion contains more bright stars than any other constellation. The hunter of Greek mythology has a three-star belt that points to nearby constellations.

Pavo, The Peacock
The stars in this constellation represent the builders of the Ship Argo.

Pegasus, The Winged Horse
The box-like body of Pegasus is a prominent feature of autumn skies.

Perseus, Andromeda's Savior
Found between Andromeda and Auriga, Perseus was the youth who saved Andromeda by slaying a Medusa and showing Cetus its severed head.

Phoenix, The Mythical Bird
Phoenix was the ancient Egyptian symbol of renewal and immortality.

Pisces, The Fishes
The fishes are connected by cords tied to their tails and joined at the star Alrisha, called "the knot".

Pisces Austrinus, The Southern Fish
Constellation near Aquarius is marked by the bright star Fomalhaut.

Pleiades
About 10° beyond the line from Orion to Alderbaran is a tight cluster of stars, the Pleiades (Plee-a-deez).

Scorpius, The Scorpion
The bright star Antares marks this fishhook-shaped constellation.

Sagittarius, The Archer
Teapot-shaped constellation is located in the vicinity of what astronomers believe to be the center of our galaxy.

The Summer Triangle
The asterism is composed of three bright stars in the southern sky in winter including Deneb (in Cygnus), Vega (in Lyra) and Altair (in Aquila).

Taurus, The Bull
Following the line of Orion's belt leads to the bright star Alderbaran, the "bull's eye" in the V-shaped constellation Taurus.

Tucana, The Toucan
Constellation near Hydrus and the Small Magellanic Cloud.

Virgo, The Virgin
The goddess of justice is marked by the bright star Spica.

Volans, The Flying Fish
Commemorates the flying fish observed by early explorers of southern oceans.

HOW TO NAVIGATE THE NIGHT SKY

Astronomers measure the distance between stars by degrees. A degree is a measurement equivalent to one ninetieth of a right angle. For example, it is about 90 degrees (90°) from the horizon to directly over your head.

The easiest way to navigate between stars is to use your hand as a celestial ruler. By holding your arm outstretched and reaching up to the sky, your hand can approximate the following measurements:

5°   10°   25°

Squint to Simplify the Night Sky

The simplest way to locate stars and constellations is to use the brightest stars to guide you through the night sky. If you are overwhelmed by the number of stars in the sky, simply squint to blot out the duller stars.

Labels: PROCYON, RIGEL KENTAURUS, ALTAIR, VELA, PYXIS, CARINA, CANOPUS, PUPPIS, CAPELLA, ARCTURUS, SIRIUS, COAL SACK, DENEB, POLLUX, CASTOR, REGULUS, VEGA, BETELGEUSE, RIGEL, ALRISHA, ANTARES, DENEB, ALTAIR, VEGA, ALDERBARAN, SPICA

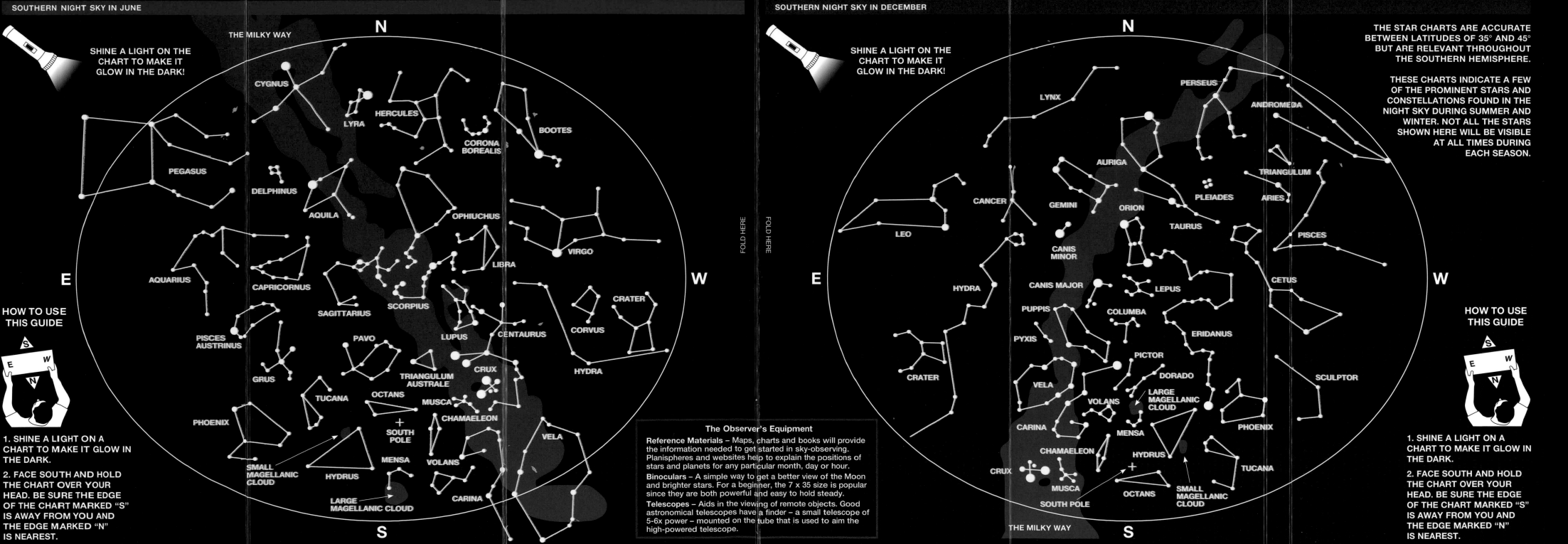

SOUTHERN NIGHT SKY IN JUNE

SOUTHERN NIGHT SKY IN DECEMBER

SHINE A LIGHT ON THE CHART TO MAKE IT GLOW IN THE DARK!

THE STAR CHARTS ARE ACCURATE BETWEEN LATITUDES OF 35° AND 45° BUT ARE RELEVANT THROUGHOUT THE SOUTHERN HEMISPHERE.

THESE CHARTS INDICATE A FEW OF THE PROMINENT STARS AND CONSTELLATIONS FOUND IN THE NIGHT SKY DURING SUMMER AND WINTER. NOT ALL THE STARS SHOWN HERE WILL BE VISIBLE AT ALL TIMES DURING EACH SEASON.

HOW TO USE THIS GUIDE

1. SHINE A LIGHT ON A CHART TO MAKE IT GLOW IN THE DARK.

2. FACE SOUTH AND HOLD THE CHART OVER YOUR HEAD. BE SURE THE EDGE OF THE CHART MARKED "S" IS AWAY FROM YOU AND THE EDGE MARKED "N" IS NEAREST.

The Observer's Equipment

Reference Materials – Maps, charts and books will provide the information needed to get started in sky-observing. Planispheres and websites help to explain the positions of stars and planets for any particular month, day or hour.

Binoculars – A simple way to get a better view of the Moon and brighter stars. For a beginner, the 7 x 35 size is popular since they are both powerful and easy to hold steady.

Telescopes – Aids in the viewing of remote objects. Good astronomical telescopes have a finder – a small telescope of 5-6x power – mounted on the tube that is used to aim the high-powered telescope.

THE MILKY WAY